Farms and Foods: New Mexico's Agricultural Legacy

Timothy

FARMS AND FOODS

NEW MEXICO'S AGRICULTURAL LEGACY

TIMOTHY MCKIMMIE

UnitySW Press
Las Cruces, New Mexico

First Edition

Published by UnitySW Ptess
Las Cruces, NM
www.unityswpress.com

McKimmie, Timothy, 1948-
Farms and foods: New Mexico's agricultural legacy/ Timothy McKimmie
ISBN: 978-0615844800
1. Agriculture -New Mexico
2. Agriculture history - New Mexico
3. Farms - New Mexico
4. Farm produce - New Mexico
5. New Mexico - History
6. Water in agriculture - New Mexico

CONTENTS

Farms and Foods is dedicated to the land and landscapes of New Mexico and to farmers and food lovers everywhere.

CLOVIS
FARMERS MARKET
4:00 PM TUESDAY
8:00 AM SATURDAY

Acknowledgements

Writers vary in the amount of help they need or get with their work. I'm not sure how I compare to other writers but I do know the value of having someone to carefully read a manuscript and give their perspective or editing advice. I've been fortunate to have several people who gave their time, expertise, and opinion regarding *Farms and Foods* and I gratefully thank them here. George Dickerson's comprehensive knowledge of New Mexico agriculture came from his distinguished career with Cooperative Extension. Kath McKimmie, my sister, is an author and journalist who, as I learned, also had editing talents. Connie Falk, a long time friend, artist, and professor of agricultural economics, is a champion of sustainable living and a former journalist. Dee Sorenson, artist and hiking partner, was no stranger to editing and helping a story flow.

I also thank the many others who generously offered their time and their views, support for the story, farm and garden tours, market and history discussions, and their enthusiasm for the many aspects of agriculture in New Mexico. Last, but not least, a remembrance of Laura "Granny" Gibson who taught me the secrets and delights of pickling, making relishes and ketchup, and other old-time farm recipes.

The Santa Fe Farmers Market

INTRODUCTION

New Mexico's farming history reaches back a long time. More than two thousand years ago Indian people grew corn, beans, and squash and learned to manipulate water for their crops. As these early cultures settled into farming and populations grew, their developing farming techniques paralleled impressive architectural achievements, evident in the grand ruins of Chaco Canyon south of Farmington. Puebloans later developed their own farming expertise along the Rio Grande floodplain and nearby foothills and mesas.

Chaco Canyon

In the early 1500s, Spanish explorers conquered the Aztecs of Mexico, then spent years building a magnificent Mexico City and expanding settlement into outlying farms and ranches. In 1598, the first Spanish colonists reached New Mexico with seeds of their crops, and livestock. Thus Puebloans came to know these new crops, and animals for food, fiber, work, and transportation. Over the years, other groups migrated and settled in New Mexico, introducing other crops and farming styles and adapting to the arid conditions.

New Mexico today reveals a land of scenic and productive farms and rangeland. Color abounds in the farmland from chile to pumpkins, from corn to peaches. The blue skies add a dazzling background.

Chile harvest - Mesilla Valley

The patterns of the land — the curving rivers, meandering canals and acequias, hay bales lying in fields, orchards of perfectly lined trees — all provide tranquility as well as a sense of stability. The green, orderly farmlands often contrast strikingly with surrounding deserts or mountains in this "Land of Enchantment."

New Mexico hay field

Photos throughout this book portray the bounty of healthy farmland and the colorful sights at farmers markets. They depict the foodsheds that connect farms, markets, dinner tables, and local communities. Readers will come to know about New Mexico's farming customs and locally grown foods.

Upper Pecos River Valley

This book begins with the foundation of agricultural history in New Mexico — how early natives developed a farming life, then the story of Spanish colonial farming, followed by westward migration of Americans to the state. The core of the book is found in chapters on vegetables, fruits, farmers markets, and farm animals. A chapter summarizes the critical relationship of water to farming in the arid climate. Later sections visit food festivals, museums, and community farms. New Mexico agriculture is chile, beans, corn, squash, and much more!

Farmers market table

Corn field and Sandia Mountains

Metate

FIRST FARMERS

The story of farming in the Southwest revolves around corn, or maize. The development of corn as a crop was a landmark event in agricultural history rivaling wheat in the Middle East and rice in Asia. Corn farming emerged in Mexico more than 5,000 years ago. Slowly developing and spreading, corn reached what is now New Mexico about 3,000 years ago. Squash and beans, too, were grown in the fields with corn. The earliest farmers in our area were likely of southwest New Mexico's Mogollon culture.

CORN

Corn petroglyph

Corn was so important because of its high calorie and energy value. It was eaten fresh, roasted, or dried, stored well, and could be carried on journeys as food or for trade. Early ears of corn only measured an inch or two but farmers selected for larger cobs, easier-to-grind kernels, and the ability to thrive under different growing conditions.

As a staple, corn cooked up many ways. Whole kernels were boiled for soups but ground corn flour was especially useful. The Indians ground corn in stone basins called "metates" using handheld stones, "manos." As the corn kernels were pounded and grated into meal or flours of varying textures, the addition of wood ash released key nutrients. Corn flours were used in cereal gruels or drinks (atole) or baked into breads and tortillas. Tamales date back as far as the Aztecs who mixed steamed corn gruel with meat or vegetables and wrapped the mixture in corn husks.

SQUASH

Squash, a second key food for early farmers, was domesticated in Mexico or Central America. The first squashes were probably used for their nutritious seeds but eventually early farmers developed a pumpkin-like squash with a tasty pulp, high in vitamins and carbohydrates. The pulp could also be sun-dried and stored for later consumption. Corn, beans, and squash complemented each other in the field. Squash leaves shade the soil, conserving moisture and reducing weeds. Bean roots provide soil nitrogen and hold the soil, while the bean plants can use the corn stalks for support.

Beans

The origin of beans as a crop was probably South America. Many kinds of beans were grown, as variable as they are today, with different colors, sizes, and cooking qualities. Beans also exhibit a large variability in growth and water requirements, thus facilitating bean selection for specific environments. Beans' protein and fiber content are two reasons for their high nutritional value. The three crops, corn, beans, and squash, provided a good balanced, stable, and healthy diet, enabling the populations of these early cultures to expand.

Bean variation

Becoming master farmers

The early farmers planted in small, widely dispersed plots around their villages and in the nearby hills. Places with good soil moisture were scattered and knowledge of local landscapes was built from years of observation. Soil moisture could be increased with simple borders of rocks or soil, and mulches held and conserved water. Such techniques were used with wild grasses and herbs, as well as with corn, beans, and squash. Carefully placed rocks might increase production of favorite berry-producing shrubs. Corn crops were quite dependent on summer monsoonal rains and corn was planted in different places to reduce the risk of crop failure. Dams of rock, earth, or brush diverted water. Fertile wetlands were planted, ephemeral canyon streams were channeled, and fields were leveled for uniform watering. Techniques varied depending on the local terrain. Canal irrigation systems were created in places like the Mimbres Valley and Chaco Canyon. Early farmers planted by the skies and seasons, but prayerful singing and dancing rituals encouraged rainfall. Crop seed was selected for regional conditions over hundreds of years, and shamans cultivated special varieties for ceremonial use.

Other Crops

In places where the growing season was long and warm, cotton was grown. The cotton fibers were used to weave comfortable clothing and blankets and the cottonseed was edible. Gourds grew into all different shapes and when dried, were used as strong, lightweight containers, utensils, and tools. Turkeys were the only livestock raised by the early farmers. Turkey feathers and down were woven into blankets and clothing.

Wild plants were gathered from near and far. Seeds and greens were cooked and eaten, others were used as medicines, or smoked. Herbs and greens like lambsquarter provided flavor and vitamins. Amaranth and sunflower offered high protein and oil content. These plants, along with some wild grasses, were encouraged in gardens or harvested from local wild stands. Mesquite pods were also collected. Some seeds were roasted, or ground in bedrock mortars that are still found today. The tiny fruits of many trees and shrubs were harvested, providing unique tastes, flavorings, dyes, and vitamins. Grasses, agaves, and yuccas were used for weaving baskets, rope and twine, even sandals.

Pottery

Pottery developed after farming arose, as people were settling into more permanent communities. Clay pots improved cooking and were important as storage vessels for dried foods. By the year 1000, an amazing variety of bowls, mugs, and eating utensils was being created from clay. Thousands of historic pottery items survive today, intact or as potsherds or shards. The elaborate painting and decoration of many of these items testifies to the artistic skills of the early farmers. Pottery also provides clues to their lives.

Mimbres bowl

Farming Communities Evolve

Ancient farming communities arose at hundreds of sites throughout New Mexico, and their ruins give us a glimpse of what existed. As the people became more adept at farming and mastery of water harvest and irrigation, their village architecture evolved into more sophisticated stone and masonry construction. Larger villages and "great houses" contained hundreds of rooms, including many for storage of surplus crops. Plazas were central features of Ancient Pueblo towns, where daily life unfolded, and kivas — underground, often circular, rooms for ceremonial gatherings — were numerous. Chaco Canyon was the center of New Mexico's world a thousand years ago; it and other large villages required vast farmlands to support them.

Chaco Canyon

While their success at farming allowed these communities to grow, some grew too large, and outpaced the ability of their environments to support them. Soil depletion reduced the land's ability to produce food. Forest resources and wild game were depleted and droughts were devastating. Today, abandoned villages are found throughout the southwest. After the year 1100, the land could no longer support the Chacoan people, forcing them to migrate or starve. Farmers carried crop seeds to new areas like Bandelier, modifying farming techniques as needed, perhaps planting on south-facing slopes to lengthen growing seasons, or in plots bordered by low walls to absorb heat. Terracing helped capture water, pebble mulches reduced evaporation, and water might be carried to crops by hand.

By 1300, many Indian farmers had moved to the fertile Rio Grande valley of central New Mexico. Here, new villages were often constructed from adobe mud. The flood plains were farmed by directing water in ditches, beginning the canal system that is still in use today. With farming success, populations grew, until there were nearly a hundred Pueblos in and around the Rio Grande valley from Taos to Socorro. Many were multi storied, with as many as 1,000 rooms. One such Pueblo near Albuquerque stretched a quarter mile along the Rio Grande, with farmlands extending out for miles, intersected by irrigation canals.

Irrigation canal

Corn varieties were developed that differed in maturation times, rooting characteristics, flour quality, taste, texture, and color. Puebloan women and girls spent hours daily, grinding and cooking corn. Puebloans were healthy and successful people with a spiritual life organized around family, farming, seasons, rain and snow, nature, and of course, corn.

Then during the mid-1500s, Pueblo communities began encountering Spanish explorers coming north from Mexico. The explorers commented in their diaries on the vastness of Indian farmlands. They called the villages pueblos, the Spanish word for "small towns." Puebloan lives would be changed forever, though many of their traditions would continue. Corn remains a vital part of Pueblo cultures and traditional foods are still prepared. Visitors can witness modern Pueblo life and farms as well as the historic ruins of yesterday.

Corn, beans, and squash

Wooden supply cart

Water-powered mill

Farm fields, northern New Mexico

FARMING CULTURES MERGE

THE 1500s

Spanish exploration in the Americas, begun in 1492, reached Mexico with Hernando Cortes in 1519. The conquest of the Aztecs was unfortunate. Aztec civilization was rich in art and architecture, and their diversity of crops nearly matched that of Spain. Many Aztec foods were to have lasting effects on the rest of the world including corn, beans, squash, potatoes, tomatoes, chocolate, and chile. While the Spanish colonists remade Mexico City, settlers spread out into valleys to grow wheat and establish sheep and cattle ranches on the immense grasslands. By 1550 exploratory expeditions reached north to New Mexico. There, along the Rio Grande, Coronado and others encountered Pueblo Indian villages with impressive farmlands. Puebloan crops were not as diverse as the Aztecs, but they had all they needed — corn, beans, and squash.

THE SPANISH SETTLE NEW MEXICO

In 1598, Spanish colonists led by Juan de Oñate settled at a beautiful and fertile area above the confluence of the Rio Grande and Rio Chama, north of present-day Española, near San Juan Pueblo. The 400 or so colonists, in fact, used Puebloan land for their new settlement and expected Indian help with their field work, irrigation projects, and living quarters. The trip from Mexico by horse, oxcart, and foot had taken six months. The colonists brought their necessities: seeds or cuttings for all their crops, livestock, flour, sugar, spices, wine, tools, millstones, church bells, and firearms. They were a mix of adventurers, soldiers, clergy, farmers, and craftsmen, hoping to start new lives while claiming territory for Spain, land for themselves, and souls for the Church.

SPANISH AGRICULTURE

The Spanish colonists found the upper Rio Grande Valley good for farming, as did the Indians. They brought their acequia system of irrigation, though Puebloans already used irrigation canals. By the second year the colonists were growing wheat, barley, and garden vegetables next to Puebloan fields of corn, beans, and squash. Their vegetables included cabbage, lettuce, onions, garlic, carrots, radish, turnips, and cucumbers, and they had a favorite new one acquired in Mexico, chile. In addition, they had peas and lentils, high protein legumes, as well as beans from the Indians. They brought seeds or cuttings of fruits such as melons, grapes, cherries, apricots, plums, and peaches. But wheat was the colonist's most important crop, since bread was their primary staple. The wheat was ground into flour using livestock or water power to turn millstones.

The Spanish meat of choice was mutton, so sheep were their most important animal, besides the horse. Their other food animals were cattle, goats, and chickens. Heavy work, such as pulling wagons and plowing fields, was done by oxen, horses, mules, and donkeys. The colonists also learned from the Indians, like when to plant and irrigate, and how to use wild plants. They were also very dependent on Puebloan corn crops and Puebloan labor. But farming the arid land was not easy and droughts increased the difficulty.

Santa Fe

The colonists needed more distance between themselves and the Pueblos, and they relocated in 1610. The new settlement, Santa Fe, was chartered by Spain as an official "villa". Santa Fe was not a bad location. There was reliable water, cooler weather, it was fine for wheat and vegetables, and the nearby mountains had resources like timber and rangelands. The new villa had a central plaza and a governor's palace where the business and legal affairs of the colony were administered. Homes were spread out, each with its own garden, and most with corrals and pasture for livestock and access to an acequia and/or spring for water.

Santa Fe was at the northern end of the Camino Real (Royal Road) that connected the colony to Mexico City. The population in Nuevo Mexico was poor and grew slowly. Santa Fe had no more than about a hundred families for the next 150 years. But other settlers chose to spread out onto ranches, or estancias, often near Indian Pueblos where they had access to water, good land, and labor. Some settled in the Española basin and nearby river valleys, while others went as far north as Taos. The Albuquerque area, especially, had plenty of good land for farms, and grasslands for livestock. The Spanish names of many New Mexico towns, rivers, and mountains originated with these early settlers. Meanwhile, church missions were being established throughout the new land.

Missions

As a partner in Spanish colonization, the church sought religious conversion. Church missions usually located at existing pueblos where there was a ready supply of souls to save. Never mind that the Indians had their own religion and priests. The missions consisted of a Franciscan friar, a church, school, living and working quarters, stables, and farmland for crops and pastures.

Isleta Pueblo
Mission Church

The missions of New Mexico stretched along the Rio Grande from Isleta to Taos, then east to Pecos, and Zuni in the west, two dozen of them by 1630. In return for religious and cultural instruction, native peoples were expected to be willing laborers and farm hands. Puebloans were taught to use draft animals, ride horses, and tend the large herds of sheep the missions raised for meat, wool, and trade. The wool was woven into blankets and other items in mission workshops, an ongoing activity requiring Puebloan labor. Crop surpluses and mutton fed the needy. Vineyards provided grapes for making sacramental wines, brandy, and vinegar. The names by which we know many of the Pueblos today, such as Santa Clara or Santo Domingo, come from the early Spanish missions. The missions themselves are gone, but a few of the Pueblo mission churches still survive.

Revolution and Reconciliation

The Indians had reasons to resent the colonists' intrusions. Pueblo land was being taken and the Indians were expected to work for the colonists, and pay tribute in blankets, hides, and crops, especially corn. The tribute system exhausted crop surpluses, created food shortages, and caused deep resentment at the meddling in Indian culture. European diseases spread among the Indians, who had no immunity. Smallpox, measles, and influenzas killed as much as half the population and many Pueblos were abandoned. Puebloan spiritual practices were suppressed and the people were sometimes treated with a punishing brutality. In 1680 Puebloans revolted, killing hundreds of colonists, and burning their houses and churches. The rest of the colonists and missionaries fled to El Paso or further south in Mexico.

Twelve years later, Spanish colonists cautiously returned to resettle the land, vowing to treat the Natives with more respect. However, the colonists demands, and reliance on Indian crops continued, to the Indians' dismay. During the 1700s settlements gradually expanded. Villa status was awarded to Santa Cruz, north of Santa Fe, and then to Albuquerque in 1706.

Trade

Life was isolated in 1700s Nuevo Mexico. The capitol of New Spain, Mexico City, was far away down the Camino Real, the trail crossing the Rio Grande at El Paso. Supplies were hauled in wooden supply carts pulled by oxen or mules. From Mexico, trade items (protected by soldiers who rode along on horses) included tools, chocolate, sugar, spices, weapons, fine fabrics, and news from the distant homeland. The colonists and Puebloans in return, sent sheep, wool blankets, and other agricultural products like wine (shipped in animal skins), vinegar, cotton blankets, and leather, as well as piñon nuts and salt. A trip on the Camino Real could take as long as six months and it might be years between wagon trains. New settlers, soldiers, and priests came north on the trail. Others decided that life in New Mexico was too harsh and returned to Mexico.

Trade at the local level between settlers, missions, Puebloans, and merchants included vegetables, fruits, grains, meat, hides, and blankets.

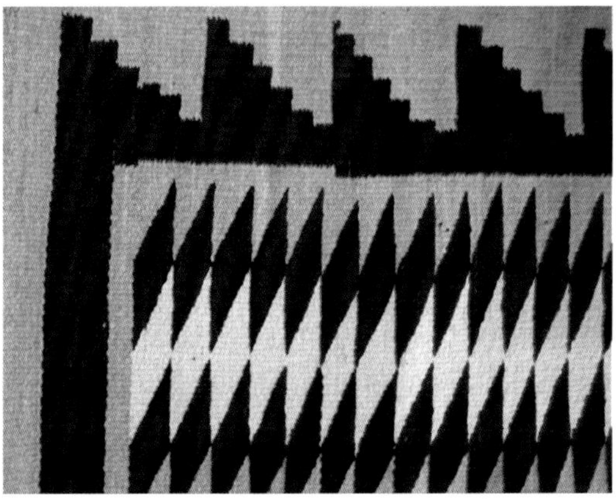

Wool blanket

Nomadic Indians including Apaches, Comanches, Kiowas, and Utes traded hides and buffalo meat for corn, squash, beans, or horses. Puebloans and Navajos traded blankets, trappers had pelts, and others traded tools, knives, and jewelry. Annual trade fairs were held in Taos and Chihuahua.

Puebloans

The Indians were skilled farmers when the Spanish arrived. Their knowledge of the land, wild foods, corn, and farming techniques supported many Pueblos. New foods like fruits, vegetables, and mutton were accepted but farming techniques clashed. Having had only two domesticated animals, dogs and turkeys, many new livestock were in their lives, sheep being the most important. Puebloans were expert weavers, creating blankets from the cotton they grew or traded for, and using local plants as dyes. At missions they trained as shepherds and shearers, and labored in weaving workshops, producing blankets for trade items. Puebloans came to favor wool over cotton for weaving, perhaps because raising sheep was easier than growing cotton. Navajos too, adopted sheep and wool, becoming master weavers. The Indian's adoption of the horse was a transportation revolution. But Puebloan populations plummeted as a result of Spanish contact. Nineteen New Mexico Pueblos survive today, about one-fifth of the number when the Spanish arrived.

New Settlements and land grants

Spanish settlers, missionaries, and Puebloans were all dependent on farming. Estancias had spread south along the Rio Grande as far as Socorro, some serving as way stations for the Camino Real, but they were hard to defend from roving bands of Apaches. New settlement was encouraged through land grants for family homesteads, estancias, or small towns. Governors could award grants for service or loyalty, sometimes including

Farming valley, northern New Mexico

rights to tribute payments from Puebloans. Other grants were awarded to petitioners who simply wanted land for a new community. The granted lands ranged from small valleys and ranches to vast tracts. Each had a source of water, land for crops, pasture, and timber, sometimes tools, seeds, livestock, and acequia or community development guidelines. The colonists' vast herds of sheep needed plenty of land to graze. As unoccupied land with water became scarce, settlements pushed into the high valleys of the Sangre de Cristo Mountains and other remote areas.

FARM LIFE - 1750

Farming in the arid land was challenging, especially in dry years. Communities on marginal lands were poor, but with luck they would get a wheat crop. The harvested wheat was spread on hard ground as sheep or goats walked over it to separate (thresh) the grain. Grinding for flour was done in mills powered by streams or livestock. Flour was stored in wooden chests called harineros. Goats were the source of milk and cheese. Diets of the poor consisted largely of beans, peas, tortillas, bread, onions, chile, fruit, and meat they raised or hunted. Chile, often the only surplus crop, was a trade item with local merchants for coffee, sugar, salt, lard, and cloth. Apples were traded for beans or chile between neighbors. Cooking was done on open fires using clay pots made by Puebloans and Apaches. Corn and wheat tortillas were cooked on flatiron or clay griddles. Baking was done in hornos, outdoor ovens made of mud. Since there was no refrigeration, fruits, vegetables, and meats were cut into strips and sun dried. Herbs were hung in bunches to dry. Horses, mules, even donkeys provided transportation.

Drying herbs

Horno oven

By 1800, settlers occupied most of the valleys of the upper and middle Rio Grande and its tributaries, and out into the Pecos and Chama river valleys. The main crops were wheat and corn, beans, squash, cabbage, chile, onions, vegetables, grapes, melons, and orchard fruits. New Mexico was remote, with a generally peaceful alliance between Spanish settlers and Puebloans, a relationship that provided some protection from disruptive raids of bands of Comanches and Apaches. Spain was focused on its California territory. Change would come with Mexican independence and the arrival of new settlers from the east.

Irrigated New Mexico farmland

Farms after 1800

By the early 1800s Albuquerque had become New Mexico's largest town. Farmers sold produce door to door from horse-drawn or donkey-drawn carts. Rio Grande Valley farms flourished, from Bernalillo to Belen. In 1821 Albuquerque and the rest of the territory became Mexican when Mexico declared independence from Spain following a protracted fight against privilege and distant rule. The Santa Fe Trail developed as a major trade route to the east. Covered wagons from Missouri brought cloth, sugar, household items, manufactured goods, and tools. Farmers and traders met the Americans on the Santa Fe plaza. Burro trains of local trade items included corn, wheat, beans, onions, chile, fruits and vegetables, eggs, milk, cheese, breads, meats, pinon nuts, and firewood. Sheep, however, were still the major export south to Mexico.

Village - northern New Mexico

Geography of 1800s settlement

Permanent settlements along the fertile Rio Grande valley extended south to Socorro by 1825. Socorro had been the site of a short-lived mission in the 1620s but farms in this area now produced corn, wheat, vegetables, and livestock. By the 1840s farming settlements reached farther south to Las Cruces and Mesilla. The first (Mexican) land grant in this area was issued to the Doña Ana colony. Settlers cleared the dense bosque flood plain, dug irrigation ditches, and planted corn, wheat, and vegetables. Farmers in these remote areas were vulnerable to attack by raiding Apaches but the Mesilla Valley was fertile and determined settlers were soon growing beans, squash, cabbage, melons, chile, onions, tomatoes, sweet potatoes, peas, lentils, and garlic. Wagons loaded with produce traveled south to El Paso, the nearest population center.

The 1840s were also a time of turmoil for the new country of Mexico. Texas had declared independence and showed interest in joining the United States. Mexico was weak without Spanish support. There were border disputes and it was difficult overseeing the northern territory, which stretched to Colorado and California. Sensing weakness and hungry for more territory, the United States launched a war and in 1848, the New Mexico Territory and more, became part of the United States.

Following the Civil War a new wave of settlers arrived and spread around the New Mexico Territory, anywhere there was water. Farms along the Pecos River soon stretched south to the Texas border. In the 1870s new irrigation canals provided Pecos River water for fields of vegetables, grains, and orchards north and south of Roswell. Artesian waters were also tapped for irrigation. Factories eventually canned tomatoes and processed sugar beets. Farther north, the Las Vegas area became prominent for wheat, flour mills, and wool processing.

Farming expanded into the San Juan, Animas, and La Plata valleys in the 1880s. Here they planted orchards and fields of vegetables, hay, and grain. Apples were especially productive and high-quality. By century's end boxcars of apples were shipping nationwide, winning awards at fairs and gaining a reputation for great taste.

San Juan Valley farms

Farmers had settled the Hatch Valley by the 1880s. Elephant Butte dam, completed in 1916 along with miles of new irrigation canals, allowed the Hatch and Mesilla valleys to become the largest crop-producing area in the state. The long, hot summers of this area were ideal for cotton and alfalfa. Other crops included chile, corn, wheat, onions, tomatoes, cantaloupe, pecans, and cabbage. Several vegetable canneries operated in Las Cruces during the 1930s and 1940s. After a major drought in the 1950s farmers drilled deep wells to support their irrigation needs. These wells allowed farming to expand to cool season crops such as lettuce and spinach. The Gila River valley was settled around 1890. Small farms there produced orchard fruits, corn, wheat, hay, and vegetables. Crops were sold in Pinos Altos or Silver City.

RANCHING

The 1800s saw a big increase in cattle ranching. The Spanish had preferred to raise sheep but New Mexico's eastern plains became part of cattle country stretching into Texas. Longhorns were the popular breed and new settlers dreamed of owning a small ranch, stories that would become part of New Mexico lore. There was also conflict between sheep and cattle owners and lives were lost. In some places battles were fought over water, too.

The late 1800s saw the introduction of windmills to pump groundwater, allowing even remote plains country to be settled. One traveler described the Deming area as a forest of windmills. With a dependable water supply, livestock could roam the vast open ranges.

Rural New Mexico windmill

By 1900 ranches had spread throughout the state. Here is how one ranch near Cloudcroft was described:

> *When the cowboys returned from the spring trail drive they hung up their spurs and got out the plows. Apples were the primary resource and were shipped; cider and vinegar were sold in the family store. Honey was also sold. Hay provided winter feed for cattle, corn for hogs and fowl. Sugar cane* (sorghum?) *was cultivated in the areas of poor soil, ground in a horse powered mill, boiled down into molasses which was also sold and was a favorite of neighboring Apaches. In addition to flowers, several vegetable gardens were planted. Almost anything that could be canned, preserved, salted down, dried, stored in the root cellar, or served fresh was planted. Any excess was sold. Hogs were butchered, hams smoked, and bacon salted. The smoke house was filled to capacity. The children did their part, the boys keeping the wood box filled, the gardens weeded, and the cows milked. Brothers and sisters tended the chickens, turkeys, ducks, and geese. The girls helped with household chores including sewing, cooking, and cleaning. Pillows and mattresses were stuffed with down from geese raised on the property.*
>
> Bidal (1995)

TECHNOLOGY AND ADAPTATION

Railroads completely changed the marketability of perishable foods. Rail lines brought settlers, equipment, and opportunities for trade, and an outlet for cattle and other agricultural products. Narrow gauge trains shipped chile, fruit, and wool to Colorado, returning with potatoes and beans. By the 1930s Route 66 and other improved roads brought truckers from Texas and Oklahoma seeking New Mexico farm products.

Advances in well pumping technology allowed water extraction from deep aquifers. New Mexico's vast eastern plains opened to cotton, beans, peanuts, and grains. Today, some areas resemble Midwestern farm country, with towering grain silos. Deep wells also opened land to farming in Deming, and in the Las Uvas, Animas, and Tularosa valleys.

Center pivot irrigation

Changing weather patterns and insects affected cropping patterns. Thousands of acres of apple trees in the lower Pecos Valley were lost to freezing weather. Codling moths appeared and ruined the fruit. By the 1930s most of that land was converted to cotton. A drought in the 1950s brought an end to most dry farming. Poverty was common but not necessarily uncomfortable. Poor farmers lacking cash could feed themselves and trade surplus crops with neighbors or local merchants. During the 1930s depression, New Mexico's farms and gardens fed the hungry and unemployed. Dust bowl refugees settled at Bosque Farms in adobe homes built by WPA workers, then started dairies, providing milk products to Albuquerque.

INDIAN FARMING

Puebloans continued their farm-based lives through the early 1900s. San Juan Pueblo, for example, was described as a great garden of fruits and vegetables. Others turned to raising cattle and sheep. Herding sheep had become a life style for Navajos whose wool blankets were highly valued. Modern life, however, reduced the acreage and diversity of pueblo farms. Corn was still grown and harvested for family use, corn dances, and harvest celebrations but hay and livestock production took over most land. More recently, with a resurgence of local foods, Pueblo farming practices have diversified, finding markets and other outlets to sell squash, chile, corn, and other agricultural products. Santa Ana Pueblo offers blue corn products at *www.cookingpost.com*. The Navajo Nation's Navajo Agricultural Products Industry (NAPI) specializes in grain crops and potatoes.

NAPI potatoes

Specialty corn

Highway commemorates Pueblo corn

FARMING TODAY

New Mexico today remains a land of many farms and diverse crops. Some land has been farmed by the same family for hundreds of years. Scenic mountain valleys are home to smalls farms featuring fruit and vegetable production at its finest and many of the state's organic farms. The acequia tradition continues, with family labor, and produce sales at local markets or roadside stands. The high value of market garden crops helps keep small farms viable.

Larger scale farms use the latest technology to produce crops such as hay and corn, pecans and cotton. All these factors provide an agricultural tradition and evolving diversity that can be enjoyed by residents, visitors, and distant markets alike.

Grain silos, eastern New Mexico

Today's most valuable farm products are from cattle, both beef and dairy. Beef, milk, and cheese account for more than 75% of state farm income. The top field crops are hay and corn, both primarily used for cattle feed. Pecans are the most valuable food crop. Number six in value is the nursery and greenhouse business, followed by onions and chile. Wheat and cotton round out New Mexico's top ten crops. With long, hot summers and few insect or disease problems, New Mexico can produce organic cotton and peanuts on a commercial scale. Other sizeable crops include potatoes, grapes, lettuce, sorghum, beans, melons, and pumpkins.

New Mexico cotton - Organ Mountains

Market table

FARMERS MARKETS

The diversity of New Mexico's fruits and vegetables can best be seen at the farmers markets. These markets are the heart of the farm and food story — big or small, they are wonderful, colorful places to visit and get fresh food. Market vendors may range from backyard gardeners to full-time farmers. The displays of fruits, vegetables, flowers, and even market signs are attractions in themselves, and markets can be cultural experiences as well. No two farmers markets are alike and each is worth a visit if you happen to be in town on a market day. Offerings differ depending on the seasons.

The Plaza tradition

Today's farmers markets continue a tradition of selling directly to the public in a central location. In historic times markets served as a town's central trading place. William Davis described the Santa Fe plaza market in his 1856 book:

> *In passing under the portal on the western end of the palace, we encounter the market on our way, where the country people sell their meats, fruits, and vegetables, they bring to town. The supply is scanty ... It consists principally of mutton, an occasional porker, red peppers, beans, onions, milk, bread, cheese, grapes, wild plums, and berries. In the winter Indians and others bring in, almost daily, fine venison and wild turkeys. Brought in on burros ... the meats are hung ... while the vegetables are put on little mats or pieces of board on the ground. Onions and beets thrive well and grow to an enormous size. The hay and grass market is on a narrow street on the corner of the plaza.*

Santa Fe plaza

It is difficult to keep track of New Mexico's farmers markets — each year there seems to be new ones. There are more than 50 markets, in every part of the state, in all sizes. Each has its own look, feel, and unique fresh food offerings and some have reputations for certain items. Santa Fe's market, the largest with more than 100 vendors, has caught the eye of national media, being called one of the top ten great places to get farm-fresh food in the country.

FARMERS MARKET SIGNS

Saturday is most commonly market day. Market rules usually state that products must be grown locally or regionally. Foods picked at the peak of ripeness often sell out rapidly, so latecomers may be disappointed. Some markets are open year-round but many are only open summer and fall. Times and locations are available at *www.farmersmarketsnm.org* or the local county extension office.

MARKET TABLES

For many market shoppers a weekly visit to the farmers market is an event not to be missed. Part of the experience is seeing local crops change with the seasons. Farmers markets also provide a community gathering place. Locals, vendors, and tourists mingle, socialize, and swap growing tips and recipes. Ready-to-eat baked goods and steamy rich coffee can be enjoyed to the sound of string quartets, guitarists, mariachis, or drummers. Special market events include apple or chile days, cooking demonstrations, food tasting, gardening workshops, health and nutrition days, harvest festivals, and even farm tours.

Flowers, herbs, honey, preserves, pickles, eggs, cheeses, meats, gourds, even aged vinegar can be found among the fruits and vegetables.

OTHER MARKET OFFERINGS

Market Grower signs

Harvest seasons

Harvest time in New Mexico varies by location. Fresh produce is seasonal, of course. Spring arrives in March or April in southern New Mexico but as late as June in the higher country. Cold frames, greenhouses, hoop houses, and row covers are ways to extend market seasons, both earlier and later.

Spring crops include lettuce, spinach, kale and other leafy greens, peas, cabbages, broccoli, asparagus, and root crops like carrots, radishes, turnips, and beets. These cool weather crops are also grown in fall.

SPRING CROPS

Summer crops include corn, squash, tomatoes, eggplant, green beans, peaches, plums, melons, cucumbers, onions, garlic, and green chile.

SUMMER CROPS

FALL CROPS

Fall markets have apples, red chile, winter squash, sweet potatoes, and pumpkins. Many spring vegetables, planted in late summer, appear again in fall.

THE ROADSIDE ALTERNATIVE

Roadside stands offer a direct way to get fresh fruits and vegetables. Often located near farm fields, these markets may only be open seasonally. Most of the produce is locally grown but sometimes crops are purchased from afar.

Vegetables

The most prominent produce at the farmers market are the vegetables. Fresh vegetables turn ordinary meals into memorable ones. Nothing compares to just-picked tomatoes, lettuce, corn, or peas. With the many choices of vegetables at farmers markets — from leafy spinach and kale to root crops like carrots, radishes, and potatoes — there's a cornucopia of choices. New Mexico's farms produce such a wide range because of the climatic diversities throughout the state, from low deserts to high mountain valleys with great variation in temperature, precipitation, and length of seasons. The history and origin of the state's vegetables are stories in themselves.

The Three Sisters

Anasazi beans

Corn, Beans, and Squash

When it comes to New Mexican food, ancient traditions rule. As discussed earlier, corn, beans, and squash all originated in the Americas and gradually moved north from Mexico along trade routes, reaching the American Southwest some 3,000 years ago. Native American myths eventually encompassed these three vegetables. In native stories the "Three Sisters" were always together.

Corn

According to archaeologists, the early farmers grew many different kinds of corn. Many historical corn varieties were lost, however, as native farmers modernized and hybrid varieties came to dominate. As the importance and value of genetic diversity became apparent, efforts at preservation of remaining native corn varieties emerged. Native Seed/SEARCH, for example, is dedicated to saving native crops. It helps grow and distribute native crop seeds, including corn, to preserve historical and heirloom varieties.

Corn remains an important crop for Indian people, and for everyone. Corn is one of the most valuable crops in New Mexico, though mostly fed to animals. Stalks are used to make silage for dairy cows, and the grain for beef cattle or mixed grain animal feeds.

Of course, people eat corn as a vegetable as well. A sign of summer, fresh roasted corn is sold at festivals, farmers markets, roadside stands. Fresh corn is also called green corn. Calabacitas is a simple, very New Mexican dish. Sliced from the cob, the corn is quickly fried with squash and onions. For year-round goodness, dried and roasted corn is cooked in soups and stews or with beans.

Corn meal, or masa, is used for tortillas or tamales. Corn flour is also used for corn bread, chips, muffins, and pancake mixes. Other corn products include corn nuts, parched corn, chicos, atole meal, grits, and hominy. Blue corn is a popular specialty item. What other crop is enjoyed in so many ways? Today, it is difficult but not impossible, to find a mill that will grind corn into meal or flour. An old mill in Cleveland, NM is open as a museum.

Beans

An essential food for early Indian farmers, beans were also adopted by Spanish settlers. There were many varieties but pintos became the favorite. By 1900 the Moriarty-Estancia area was known as the "Pinto Bean Capitol" and New Mexico led U.S. production. Beans were dry farmed, that is grown solely with rainfall, until a major drought in the 1950s. Today, pintos and other dried beans like bolitas are usually irrigated. Harvest is in late fall. Peas and peanuts are bean relatives.

Squash

Squash was first grown for the nutritious seeds, or pepitas. Even today, roasted squash or pumpkin seeds are a treat. When squashes were later selected for their thicker and tastier flesh, they were eaten fresh, or dried for later use. Squash can be selected for numerous characteristics and many different kinds are available today. Summer squashes include zucchini and other tender squashes like yellow-crookneck. Hard-shelled squashes such as acorn, butternut, hubbard, and pumpkins are available in fall and store well for winter use. Excess pumpkins even make good fodder for livestock.

CHILE

Small round seed pods on delicate pepper plants found in Central America were originally harvested as a wild spice. Over the centuries the chile's use and cultivation evolved, until eventually a mix of chile, chocolate, and peanuts became mole, a dish known to the Aztecs. Peppers traveled north from Mexico somewhat later than the "three sister" crops but quickly found their way into the southwest culture. Chile was a perfect plant. It added spice to dull meals and kept for long periods after it turned red and dried. Chile was an accepted currency in the early days of New Mexico. A prolific crop, each plant can produce a hundred pods.

Green chile

New Mexico remains an excellent place to grow chile and is world renowned for chile diversity and flavor. The harvest begins in mid-July with the much-anticipated picking and roasting of the first green chiles. The harvest continues throughout summer into fall as green chiles show up statewide at markets, farm stands, and trucks parked along well-traveled roads. Standard varieties of green chile include Big Jim, Joe Parker, and Sandia and chile may simply be sold as medium, hot, extra hot, and so on.

Roasting chile burns off the skins, adding an appealing complexity of flavor of caramelized sugars, chile heat, and roasted-burn taste. In late summer the spicy aroma of roasted chile is everywhere in New Mexico. Roasted green chile is used in salsas, enchiladas, rellenos, soups and stews, breakfast platters, and burgers. Frozen or canned, green chiles supplement meals throughout the year.

Green chiles eventually turn red as the pods begin to dry on the plant. Red chiles are often strung as a bunch into ristras for storage, but many shoppers buy them for wall or door decorations. Dried pods keep a long time. Red chile paste or sauce is the prime ingredient of many New Mexico dishes like enchiladas. Dried red chile is also crushed into flakes or ground into powders. Commercially packaged chile powders and spice mixes are sold worldwide.

Stringing a ristra

Chile tops the menus of New Mexico, where more chile is eaten per capita than in any other state. Chile production increased with the growing popularity of Mexican food across the country in the 1960s. New Mexico is the No. 1 state in chile production, chile processing, and chile research. The value of the chile crop is comparable to onions, with both crops trailing pecans. New Mexico celebrates chile in many ways, such as a chile festival in Hatch, whole enchilada fiesta in Las Cruces, and fiery foods show in Albuquerque.

Fall color

Though chile grows up to 7,000 feet elevation, most commercial production takes place in the warm climate and long growing season of southern New Mexico. In many small communities like Hatch, chile sales to locals and tourists provide an important income. But most of the chile crop is processed commercially at plants in Las Cruces and Deming.

Chile varieties in some northern New Mexico valleys date back several hundred years. Traditional varieties have unique flavors or disease resistance; some are even said to be frost resistant. Traditional chiles may make up for their smaller sizes with their "terroir" allure. Chile harvests in the north are relatively small. The peppers are usually available only at local markets and roadside stands.

Many varieties of chile peppers grow in New Mexico besides long-green.

Paprika is a pepper that lacks heat and is grown for its oleoresins, which are used as red or orange coloring agents in foods and cosmetics. Paprika is a big industry in Spain, but New Mexico grows most of the U.S. paprika crop.

Cayenne peppers are used for hot sauces, first mashed then combined with salt and vinegar for aging and fermentation. These peppers are also dehydrated for red pepper powders and spice mixtures.

Jalapeño peppers spice up any dull meal and are sold fresh for salsas and cooking. Commercially, jalapeños are processed into sauces and dehydrated for spice mixes. Smoke-dried jalapeños are called chipotle and have a hearty, exotic taste.

Jalapenos

More Vegetables

The cabbage family represents the largest group of vegetables and includes cabbage, mustard greens, collards, kale, broccoli, cauliflower, Brussels sprouts, kohlrabi, bok choy and other Chinese greens, turnips, radishes, and arugula. Cabbage was very important to Spanish settlers and it is still the most widely grown vegetable of this family. It is a versatile food eaten cooked, raw, or fermented into sauerkraut, it stores well, and the excess can be fed to livestock. Cabbage family vegetables are grown in cool seasons because hot weather causes a strong flavor.

Turnip greens

Cabbage field

Carrots were originally cultivated in the Middle East and like many vegetables, were brought as seed from Spain to Mexico in the 1500s, then to New Mexico. Early carrots were purple or white. Carrots were a good cash crop in New Mexico in the 1930s and 1940s, shipping out of state by the boxcar from the Bluewater-Grants area. Today many different shapes, colors, and varieties of carrots are found at farmers markets during spring, summer, and fall.

Cucumbers were originally cultivated in Africa and came to New Mexico with Spanish settlers. Different kinds of cucumbers are grown for fresh eating or pickling. Pickle factories emerged in New Mexico from time to time but never succeeded for long.

Green beans are the same species as pinto beans, a reflection of the variability of the bean family. Green beans are popular in family gardens and at farmers markets. The many varieties differ in sizes, textures, and colors, such as yellow or purple "green beans." The harvest season is long, from early summer to late fall. Green beans were a big crop in the 1940s and 1950s with canneries in eastern New Mexico. At their best as a fresh backyard garden crop, green beans can also be frozen.

Lettuce is one of the oldest of vegetables. Romaine lettuce was savored long ago by none other than the early Romans. A cool season crop, lettuce was brought to New Mexico by Spanish settlers. Some of the finest lettuce is grown in the mild climates of northern New Mexico. These tender, tasty, and colorful leaves are popular at farmers markets. Winter-spring lettuce can be grown commercially in southern New Mexico, where it is also one of the finest crops for winter-spring gardens.

Red leaf lettuce

Spinach is eaten both raw and cooked. Grown commercially on about 1,000 acres in the state, spinach farming is labor intensive but can be profitable. Consumers anticipate the harvest and don't mind paying a premium. Spinach is a spring and fall crop in southern counties but a summer crop in the north.

Spinach field

Chard and beets are spinach relatives. They seem to be easier to grow than spinach but require a longer season. They are very similar to each other but chard is usually cooked as a green. Beet leaves are also cooked but the round red (or orange) roots are boiled, baked, or pickled into tender delicacies.

Beets or chard?

A cooking staple, onions were just as important for Spanish settlers as they are today. Mesilla Valley onions once grew huge, up to 7 or 8 inches in diameter, too large for today's consumers. Onion farming has expanded to compete with chile for the state's most valuable vegetable. Onions are commercially grown in southern counties, harvested in spring or summer and sold fresh, processed in sauces, or dried as a spice. In Las Cruces, New Mexico State University's onion research program breeds onions for taste, color, size, shelf life, disease resistance, and harvest time.

Garlic is an onion relative, grown in small quantities around the state. Garlic takes nearly nine months to mature with harvest beginning in May and continuing through the summer. Local garlic is found at farmers markets, often in decorative braids.

Potatoes, native to South America, were a popular crop in New Mexico by the early 1800s. It is not clear when they arrived but apparently not with the early colonists. Today's crops are mostly sold as "chippers" for potato chips, and grown in eastern counties, as well as by Navajo Agricultural Products (NAPI) near Farmington. But distinctive specialty potatoes show up at markets, especially in northern counties. New potatoes are usually dug in spring, while the bigger ones make the summer harvest.

Tomatoes are also from Central America though they were not brought north with the early colonists. Garden-fresh tomatoes are summer favorites, big sellers at farmers markets, and used in salsas, salads, sandwiches, sauces, soups, and stews. New Mexico tomato farms once supported canneries but today's tomato market is local. The prices and taste of some "heirloom" varieties such as Brandywine may inspire backyard gardeners to plant their own. Tomatillos are a relative of the tomato, and though not widely available, their unique taste makes them popular for salsas.

Eggplant is originally from Asia. It has many varieties, sizes, colors, and shapes. An early variety was small, round, and white, hence the name. Eggplant is in the same family as tomato, chile, and potato. Eggplant is another versatile addition for summer cooking. Fresh eggplant, squash, tomatoes, chile, and onions make the start of a great meal.

Other vegetables of all kinds are grown in gardens around the state. Specialty vegetables and herbs such as asparagus, basil, bell peppers, bok choy, cilantro, okra, parsley, peas, and sweet potatoes are a treat to find at local markets.

Fresh peas

Black-eyed peas

Basil

NEW MEXICO GARDEN VEGETABLES

Yellow crookneck squash

Okra

Sweet potatoes

Brilliantly colored bell peppers

Apple orchard - central New Mexico

New Mexico cherries

Fruits and Nuts

Early on, the Spanish colonists introduced many of the fruits we enjoy today — peaches, apricots, cherries, plums, apples, melons, grapes, and a little later, apples. These cultivated fruits were unknown to the Indians though they harvested tasty and nutritious fruits of many plants such as wild plums, berries from shrubs, yuccas, and the fruits of some cacti.

New Mexico plums

Fruit harvests today highlight summer and fall, and are key seasonal components of farmers markets. Many fruit crops thrive in New Mexico's climate, with irrigation, of course. On the other hand, harvests are often reduced by weather extremes — hail, freezes, drought, even wildfire — as well as birds and other wildlife. In good years, however, an abundance of fruit produces enough to can or make jams and jellies, cider and applesauce, pies and fruit wines.

Nuts are a different story. Indians have long gathered and roasted piñon nuts, an important seasonal component of their diet. But pecans, peanuts, and pistachios would not be grown in New Mexico until after the American settlers arrived.

New Mexico pecans

Fruit Orchards

Orchards were an important component of New Mexico farms for Spanish colonists and later for American settlers. Historically, orchards comprised a wide variety of fruit trees. It was not unusual to have 40 different kinds of apples, for example, or a dozen kinds of peaches or plums. Each variety had slightly different characteristics for cooking, fresh eating, and cider, and they had different maturation dates. Orchards are not only less common today, they are less diverse, so harvest seasons are shorter. One week there is an abundance of fresh ripe fruit, the next it may be gone. Most fruit sales are local and fruit lovers anticipate the seasons.

Apple blossoms

Apricot orchard

Apples are the state's most common orchard fruit. Often propagated by grafting, a few varieties growing today may have originated as trees planted by Spanish settlers. By 1900 New Mexico apple crops were large enough to ship to other states by rail and truck. Roswell apples won first place in a 1901 international exposition. But apple production declined in the 1940s due to a combination of insect pests, weather, and competition from other states, especially Washington. Still, apples are New Mexico's most valuable fruit, though sales today are mostly confined to local or regional markets.

In some years apple crops are completely lost to late spring frosts. The trees are usually sprayed against codling moths, the parent of the worm in the apple. In years when all the conditions are right, apples saturate the market bringing prices down, a deal for shoppers. Some orchards allow customers to pick their own fruit.

The bounty of sweetness from orchards lasts from late spring through fall. Apricots usually ripen first. Peaches may be harvested early or late depending on the variety. Plums are desirable for their taste and for their vibrant colors. Cherries are a rarity, requiring higher elevations, and fetching higher prices. Pears may also be hard to find. They, along with apples, are harvested last.

New Mexico pears

New Mexico peaches

Baskets of market fruits

More Fruits

Casaba melon ready for picking

Melons of all kinds do well in the hot New Mexico summers. Melons originated in Africa and Asia. Watermelons are a perennial favorite but other popular melons include the netted-rind cantaloupes and the smooth-rind casabas and honeydews. Melons are picked from mid-summer through fall.

Berries are specialty crops with high value but fairly short harvest seasons. Most common are strawberries, blackberries, and raspberries. Some berry farms offer "you-pick" opportunities with gift shops that sell jams and ice creams. Lately pomegranates are getting more attention and respect. The juice is said to be a health tonic. Fig trees are grown in southern New Mexico although they may freeze to the ground in winter.

Pomegranate juice

Raspberry harvest

Grapes and Wine

New Mexico's first vineyards were established by Franciscan friars who accompanied Spanish colonists. The first wine was likely produced in the 1630s at the mission at Senecu Pueblo near Socorro. Wine was an essential part of church ritual and the New Mexico climate was good for grape growing. Wine from mission vineyards was eventually shipped to Mexico on the Camino Real. Wine production peaked in the 1880s, when expansive vineyards lined the Rio Grande valley around Albuquerque, Socorro, and the Mesilla Valley. Various factors including floods, freezes, drainage problems, pests, and not least, competition from California wines, caused New Mexico's wine industry to collapse by the early 1900s.

Vineyard - southern New Mexico

More recently, since the 1980s, New Mexico's wine industry has rebounded. Increased consumer demand, new, more select grape cultivars, and more knowledgeable winemakers have improved wine quality. New Mexico wineries produce some award winners and wine demand continues to grow.

Some 50 wineries are located around the state, many with tasting and sales rooms. At wine festivals, nearly a dozen of them, the public can compare wines from many different New Mexico wineries. A wine trail map can be found at *www.nmwine.com*.

Vineyard - northern New Mexico

Nuts

Four nuts comprise New Mexico's nut industry, though one is not cultivated but is wild.

Pecan orchard - Mesilla Valley

Pecans are native to the southeastern United States where they were enjoyed long ago by Indians there. Pecans were grown commercially by the 1800s. New Mexico pecan production began in the 1940s and acreage continues to increase today. They grow well in hot, arid parts of southern New Mexico, but they require more water than other crops. Pecan orchards can be scenic, especially the shady overhangs and fall color. They are harvested in December and January. Pecans are the most valuable food crop in the state, worth more than $100 million annually. New Mexico vies with Georgia as the leading pecan producing state.

Originally from South America, peanuts are known as nuts though they are relatives of beans and peas. Peanuts, also called groundnuts, grow underground. Peanut acreage in eastern New Mexico is expansive and growing. The relatively dry conditions there minimize diseases so organic production is possible. Demand for organic peanuts and peanut butter is high and the peanut hay, the leafy above ground part of the plant, can be baled and sold to organic dairies.

Pistachios are originally from the Middle East and were being grown in Spain by the first century. They do well in hot dry areas like New Mexico's Tularosa Valley. Pistachios are a popular and valuable nut and their sales rooms have become tourist destinations.

Pistachio trees - Tularosa Valley

Piñon nuts, the state's indigenous nut, grow wild in the hills. Early native peoples all over the southwest roasted these nuts, harvested from a pine species, *Pinus edulis*. Commercial harvests of New Mexico piñon once shipped worldwide. Generally, there are bumper crops every few years, ranging across New Mexico. A favorite for many New Mexicans, piñon nuts are often available raw or roasted from trucks along roadsides, after the fall harvest.

Pinon cone and roasted pinon nuts

Wildflowers: eastern New Mexico plains

Gallardia with bumblebee

New Mexico-grown roses

Drying flowers

FLOWERS AND SPECIALTY CROPS

Flowers provide us with spectacular colors, patterns, and beauty, while attracting birds, bees, and butterflies. With adequate rainfall, wildflower meadows bloom in many areas throughout the state. Homeowners and businesses enhance their landscapes with flower plantings and growers have them available at farmers markets and nurseries.

Many kinds of flowers thrive with New Mexico's warm days and cool nights. Flower growers cultivate their favorites for display in enticing bunches at market stalls. Some sell as cut flowers like zinnia, sunflowers, carnations, and sweet peas. Others, such as statice or lavender, are used in dried flower arrangements, sometimes combined with greenery from juniper or sage.

Flower arrangements
Santa Fe farmers market

A few showy fields of flowers are open to the public. Lavender, with its scenic purple fields, is attractive to visitors. Lavender festivals, held at the peak of the summer season, offer soaps, sachets, dried arrangements, and lavender oils.

New Mexico lavender field

New Mexico has a sizeable greenhouse and nursery industry. Many plants are raised, including landscape trees and shrubs, annual bedding plants, cacti, succulents, natives, house plants and holiday plants, and cut flowers like roses and carnations.

Potted nursery plants

Native plants at nursery, ready for transplant

Honey is a product of flowers and is a valuable "crop", popular at farmers' markets. Honey is a natural sugar with distinctive flavor depending on the flower sources. Just as important, the bees responsible for gathering the flower nectar, and making and storing the honey, are also essential crop pollinators.

Herbs thrive heartily in New Mexico's high, dry climate. With increased interest in herbs in recent years, dozens of culinary, medicinal, and dye-plant herbs are available in local markets, specialty stores, and over the Internet. New Mexico State University's Sustainable Agriculture Center in Alcalde is a research center for herb cultivation.

New Mexico herb garden

Happy horses - northern New Mexico pasture

Criollos

Longhorn

LIVESTOCK AND RANGELAND

Following a familiar pattern, most of the farm animals we know today, cows and horses, mules, donkeys, sheep, goats, even pigs and chickens were introduced by Spanish colonists. Turkeys, however, are native to New Mexico and were raised by Ancient Puebloans. For Spanish colonists, lamb or mutton was the preferred meat. Sheep were easier to herd than cattle, are prolific breeders, and their wool was very important. Sheep were New Mexico's dominant livestock for 300 years. Spanish cattle, the Criollo and Longhorns, are still raised but today's most common cattle breeds are Angus and Hereford, introduced in the late 1800s. Since those days, cattle ranching became supreme, and cowboys, cattle drives, ranches, open range, and even range wars became imbedded in state history.

New Mexico rangeland produces the grasses and other plants on which livestock thrive. Rangeland, whether private, state, tribal, Bureau of Land Management, or National Forest, is the largest agricultural land use in the state. Some of the finest grasslands are in New Mexico's northeast plains, including Kiowa National Grassland. Ranches comprise the largest portion (90%) of New Mexico's farm acreage so it is no surprise that livestock, primarily cattle, dominate state agricultural income.

Beef cattle - northern New Mexico rangeland

Cattle

Milk cows and beef cattle account for three-fourths of state agricultural receipts. Dairies thrive in the warm, dry climate of southern and eastern counties, where disease problems are minimized. Cheese factories, a relatively new industry, convert most of the milk into mozzarella, often used for pizzas.

Beef cattle first began to dominate New Mexico rangeland after the Civil War. The range was open and cross-county cattle drives, a symbol of the west, moved the cattle to market. Today's rancher depends on fencing but great expanses of rangeland are still filled with cattle. A beef cow and calf can spend much of the year grazing the range. Beef from local ranches can sometimes be found at farmers markets or specialty grocers.

New Mexico beef cattle

Beyond rangeland grazing, most cattle feed is produced on regional farms. Hay and corn are the major field crops, both grown to feed cattle. Alfalfa is the most widely produced hay, but grass and mixed hays are also grown. Chopped and fermented corn stalks — silage — is a primary feed for dairy cows. Grain corn, on the other hand, is a major component of feedlot diets, as well as in other animal feeds.

Corn - southern New Mexico

Grass hay - southern New Mexico

Other Livestock

Though sheep numbers are reduced, wool still supplies a small industry of weavers throughout the state. In Mora, a wool processing plant cleans, washes, cards, and spins the wool fibers. Blankets and rugs, shawls and scarves are available at specialty shops. Navajos still raise herds of sheep; their wool blankets and rugs are prized around the world. Goats provide meat and fiber (mohair or cashmere). Goat cheeses such as feta can be found at farmers markets.

Weaving shop

New Mexico buffalo

Horses entered state history with conquistadores who rode in on horses covered with armor. Indians soon learned the advantage of horse travel, which also enhanced the raiding activities of some tribes. Later, horses became essential for herding cattle. Today, horses are found statewide. Ranches and farms use them for herding and fieldwork. Others use them for pack animals, recreation, breeding, in rodeos, horse shows, and as pets. A small racing industry supports several racetracks. Mules and donkeys are raised for work, packing, or enjoyment. Alpacas and llamas are kept as pets, used as pack animals, and sometimes for their fibers. Bison, or buffalo, brought back from the brink of extinction, once again roam the range. Chickens and turkeys are raised in small operations throughout the state for meat and eggs. Other fowl include ducks, geese, emus, and ostriches. A few farms raise hogs or rabbits.

Free-range turkeys

Though New Mexico rangeland has a history of being severely overgrazed, it remains important for livestock as well as wildlife. Big game animals include deer, elk, bear, pronghorn antelope, mountain lion, and bighorn sheep.

New Mexico rangeland

Farmland, acequia, Sandia Mountains

WHAT ABOUT WATER?

Farming arid lands requires an intimate knowledge of water. Snowmelt, summer storms, springs, even hillsides provide myriad water collection points that can be planted. Observation of patterns of soil moisture and precipitation led Ancestral Puebloans, over a period of several thousand years, to a mastery of farming and eventually to canal irrigation techniques similar to those we use today. Spanish colonists brought their own traditions of water management, not unlike the Puebloans', but more formalized. Spain is a dry land, requiring supplemental irrigation, similar to New Mexico. The Spanish acequia system, developed over centuries of farming, has special rules for cooperative water management.

Quiet acequia

Acequia irrigation canals have served many New Mexico farms since 1598, anywhere a stream could provide reliable water. Windmills in the 1880s gave access to harder to reach, undergound water. During the twentieth century, dams and other large-scale water projects boosted agricultural production, although with the water supply nearly always limited, cooperation remained essential. As this book goes to press New Mexico is in its worst drought in recorded history. Water has become a severely limiting factor and farming is now forced, for the time being at least, to shrink.

Irrigated farmland - central New Mexico

ACEQUIAS

Acequias are common features in the landscape of northern New Mexico. They often resemble streams lined with trees, shrubs, and grasses. The term "acequia" can indicate a single canal or an entire community irrigation system. The main or mother ditch is the acequia madre. From the acequia madre, water flows through a system of gates to smaller lateral ditches and eventually to individual farm fields.

Owners of farmland along acequias are known as "parciantes." The rules for water use and apportionment are overseen by a "mayordomo." He or she is responsible for inspection, cleaning, and maintenance of the acequia madre, settling water disputes, and general accounting. Spring acequia cleaning may take days as parciantes provide labor for removal of mud, overgrowth, debris, and trash so the water can flow freely. Acequias link all the farms in the community.

New Mexico acequias

Acequia communities help supply an increasing demand for fresh and organic foods while ensuring economic opportunities for rural New Mexicans. They represent resource sharing, an identity, and a common culture. New Mexico's acequias connect the land, food, and beauty of the state.

Increased water demands, whether from population growth or industrial development, are ongoing. Acequia communities and traditions are under pressure to develop the water and farmlands. Some call for lining ditches with concrete or pipes to save water. But that course would be slow death for trees, shrubs, and wildlife that depend on acequia seepage and habitat. Author/farmer Stanley Crawford (2003) explains:

The acequia, I suspect, is shocked that anybody might consider this a somehow wasteful arrangement. Leaks are what the acequia is all about. It creates countless habitat corridors out into the fringes of our river valleys, right into our backyards, and it will charge and recharge the aquifers beneath our feet.

Some local communities are organizing to preserve their acequias and farms. Spring cleaning on some acequias has become a celebration. The National Historic Preservation Act (1966) recognizes that such living parts of communities as acequias, are worthy of preservation. Crawford (2003) once again sheds light:

(Acequias) are repositories of a unique tradition of rural self-governance with few parallels elsewhere in rural America. As gravity-fed irrigation systems they are simple, energy efficient, and ecologically sound, and their hand dug watercourses serve to expand the riparian habitats from which they flow. As systems serving small parcels of arable land, they have proven historically sustainable over centuries in a way that grander irrigation works never have and perhaps never will, anywhere in the world. And because they make few claims on the public works budgets compared to other features of the rural infrastructure, employing local, indeed neighborhood labor almost exclusively, they may eventually prove to be valuable models for future small scale, self-governing institutions in rural areas. Acequias are one of the central institutions of Hispanic culture in northern New Mexico and their extinction would be an irrecoverable cultural loss.

Inspecting a small acequia

WATERSHEDS

Watersheds are important in many ways. Mountains store water before releasing it in streams and rivers. The water is renewable — for drinking, farming, fish, forests wildlife, rangeland, homes, and industry. Spring runoff, which provides water for farms, is predicted by measuring winter snow pack.

New Mexico's Sangre de Cristo Mountains

The watershed concept is useful in describing the limits and boundaries of a water supply in arid regions. Watersheds can be local or regional and include deserts. A watershed might be considered as one whole side of a mountain range and its streams. The Sangre de Cristo Range, the highest in the state, with 13,000 foot peaks, holds the most water, and multiple watersheds. Stream and river health are a function of watershed health and special status may be granted to some rivers in order to protect them.

Thunderstorm in the Sangre de Cristos

The Rio Grande

The Rio Grande is New Mexico's biggest river. It originates in the mountains of Colorado, as do the Chama, Animas, and San Juan rivers.

The Rio Grande runs the length of the state. Entering New Mexico near Questa, it absorbs many tributaries such as the largest, the Chama River. The Chama has three dams: Heron, El Vado, and Abiquiu. The Chama joins the Rio Grande north of Española, an area historically important for pueblo locations and the first Spanish settlement.

The Rio Grande - northern New Mexico

Chama River

Cochiti dam impounds the Rio Grande west of Santa Fe. The Jemez River from the Jemez Mountains joins the Rio Grande north of Albuquerque. Smaller Rio Grande tributaries further south include the Rio Puerco and Rio Salado and streams draining from the Sandia, Manzano, Magdalena, San Mateo, and Black ranges.

Abiquiu Reservoir

Near Truth or Consequences the Rio Grande has two dams, forming Elephant Butte and Caballo reservoirs. These provide storage, flood control, and irrigation for hundreds of farms in the Mesilla and Hatch (Rincon) Valleys.

Irrigation - Mesilla Valley, New Mexico

South of Caballo Reservoir the Rio Grande has become a part-time river, flowing only when irrigation water is delivered to farms, from March to September if water is plentiful, or only for a month or two in midsummer during years of water shortage. The river here has been straightened, looking like a big canal. Farms fill the Mesilla Valley all the way to El Paso where the Rio Grande becomes the international border with Mexico.

The Rio Grande - southern New Mexico

Other rivers

The Pecos and Canadian rivers flow from the south and east sides of the Sangre de Cristo Mountains. The Pecos begins in rugged canyon country with only a few farms in small valleys and river bends. The river valley widens at Fort Sumner supporting larger farms. Farther south near Roswell, the Pecos along with Sacramento Mountain tributaries irrigate a major farming region. The Pecos then flows into Texas. The Canadian River, flowing east from the Sangre de Cristos, is dammed at Conchas and Ute reservoirs, irrigating farms in northeast New Mexico, then flowing east to meet the Arkansas River in Oklahoma.

Pecos River - central New Mexico

The San Juan River joins the La Plata and Animas Rivers near Farmington. As with other areas, the irrigation water here has turned valleys green with crops and orchards. Nearby, the Navajo Agricultural Products Industry (NAPI) irrigates mesa farmlands. The San Juan River eventually joins the Colorado River in Utah.

San Juan River

In southwest New Mexico the Gila River rises in the Mogollon Mountains. This rugged area has limited farm acreage. The Gila is one of the few remaining perennial streams in the Chihuahuan Desert. The riparian habitat, flowing through the desert, is critical for many plants and animals. The Gila River flows west through Arizona where it joins the Colorado River. The Mimbres River, flowing from the Black Range, supports a small farming valley before disappearing underground into the aquifer.

Gila River

WATER SUPPLY AND MANAGEMENT

The drought gripping the state in 2013 intensified water rights claims and distribution problems. It is difficult to predict what will happen if it continues. Weather is fickle, water supply may not meet demand one year, while flooding can destroy crops the next. The State Engineer is responsible for water management, a juggling act that must consider the needs of New Mexico's cities, homes, and farms as well as other states and nations' rights to water.

A dry Rio Grande

During wet periods, dams and reservoirs can store enough water to last several years, a water banking system. The drought of the early 21st century has now emptied them, rivers have gone dry, and water to farms has been curtailed. Unless groundwater is available, farmland is left unplanted (fallowed). There are few options for long-term drought. More drought-tolerant crops or water conservation methods like drip irrigation may be helpful.

Dam for water storage

Farms with access to groundwater have advantages. Groundwater aquifers are like underground watersheds or underground lakes. The water is used for irrigation by pumping to canals, sprinklers, or drip systems. Aquifers are renewable sources of water if their recharge rate is adequate, but many are shrinking, some at alarming rates. At greater depths, pumping costs increase or worse, the water may get too salty or run out altogether. New wells can be drilled, if permitted, but that might reduce water supply to others. It gets complicated. Water transfer is a new challenge. For example, a pipeline could transfer water from farmland to city. What happens to farmland without water?

Irrigated farms can also have problems with too much water or with salts. If soil drainage is not adequate, flooded roots can kill crops. Few crops can tolerate saline conditions so salt buildup is managed by applying extra water. Large, open drains, not to be confused with irrigation ditches, are common in areas like the Mesilla Valley. They serve to prevent high water tables and to move salts away.

Water is a commodity that is becoming more expensive as a result of growing cities, towns, and new industry. Acquisition of farmland, with its associated water rights, is one way to obtain a water supply. However, a region's ability to produce adequate and affordable supplies of good food (food security) may be compromised as farmland disappears. A debate is evolving over whether to make deliberate decisions to conserve farms and water rights. Some farmlands have been permanently protected by easements and trusts. Decisions on how best to use water are especially challenging in times of drought.

Stream flow from Pecos Wilderness

Community garden - Albuquerque

Central New Mexico CSA

Community Gardens, Organic Farms

Communities provide many opportunities for people to get involved with food production. Those without gardening space of their own can often find and join a community garden. In addition, there are ways to join group efforts that support local farms while enjoying the taste and diversity of local foods and sometimes even participating in crop harvests.

Community gardens are usually located on city or county land but some local gardens may be shared on private land. Members pay a moderate fee to use a small garden plot including water, and sometimes hoses and tools. Gardeners working together can share ideas and expertise, learn about the growing seasons, and trade excess produce. County extension offices may provide locations of community gardens.

At one community garden, the Rio Grande Community Farm in Albuquerque, about 70 garden plots are available. In one area, elementary school children plant a garden for wildlife, then study the plants, birds, butterflies, and other animals it attracts. An adjacent farm, Los Poblanos Fields, grows traditional crops while maintaining open space and providing habitat for urban wildlife. In the fall they cut a corn maze.

In the Las Cruces area, *La Semilla* (The Seed) *Food Center* works with young people and schools. They grow organic foods and learn the basics of nutrition, food distribution, and marketing. Some communities partner with local farmers via food banks or school lunch programs. *Farm to Table* is a statewide organization dedicated to sustainable agriculture and connecting local foods to local markets. Some communities and neighborhoods have gardening and sustainable living fairs. Earth Day celebrations may also highlight gardening and food production. Seed exchanges are simple ways for gardeners and community members to gather and share vegetable and flower seed at little or no cost.

Supporting local farms keeps them viable, helping protect them from development, while increasing food security and community health. Farms that grow for local markets produce a greater diversity of crops, leading to a longer harvest season. Restaurants, school districts, food processors, and consumers that purchase locally grown food keep those dollars in the community.

Community Supported Agriculture (CSA)

CSAs are basically agreements between farmers and local consumers. They enable consumers to get the fresh food they want while supporting or even participating on a local farm. In a typical arrangement, a CSA membership for a family costs around $500 a season or about $20 a week. CSAs vary from 30 to 300 participants. Each week members receive fresh-picked fruits and vegetables, sometimes including eggs, meats, herbs, or flowers. Most CSAs farm organically. Seasons vary, from 20 weeks to nearly year-round.

CSA members share farming risks like weather, insects, and crop disease with farmer(s). In good years members receive armloads of produce, sometimes more than they can eat. However, if crops fail members receive little. Farmers benefit by having a steady income and reducing the risk of production fluctuations. The system encourages careful soil building and crop tending as well as biodiversity. Growing seasons are extended by planting a wide variety of vegetables in small plots with staggered planting dates.

CSAs in Spring

At some CSAs members can participate in farm activities like weeding, harvest, distribution, or newsletter production. About fifty CSA groups farm in New Mexico, mostly near Albuquerque and Santa Fe.

Organic production

Organic production is first and foremost, about healthy soil. Natural, renewable materials such as composts and manures help build soil, retain water, and provide plant nutrients. Using natural pest control and fertilizers is earth-friendly. Weeds are managed by tilling, pulling, and mulching instead of synthetic herbicides which have potential problems including health risks. Animals raised organically get no hormones, and antibiotics only if they become ill. Organic farming is not new; most synthetic pesticides have been around only since the 1940s.

Organic farms seem to produce exceptional food. Healthy crops result from good soil and careful hands-on tending. Farmers must find innovative solutions to pest problems and crop choice may depend on soil conditions. Organic growing is not the solution everywhere but it is a worthwhile condition to strive for. If weeds or other pests are pervasive, often from poor prior land management, organic farming may be difficult or impossible.

Organic farm central New Mexico

Standards for organic food were developed by the USDA for consistency and reliability. Food that is labeled "organic" must be certified. Only a small fraction of New Mexico farms are certified organic. Many farmers, however, use organic methods without certification status. Some may reject the required record keeping. Other farmers prefer a higher standard and stricter requirements than certified organic in matters such as freshness, local production, and small scale. Hands on pest control is possible with small scale agriculture. Growers who do not follow strict organic practices often forego the use of synthetic pesticides. They may display "pesticide-free" signs at farmers markets. For many consumers, pesticide use on foods is a primary concern. Pesticides pose serious health risks to farmers, consumers, wildlife, and the land. Pesticide resistance is also a problem, as pests become tolerant to the chemicals.

Crop diversity, growing a wide variety of crops, is important in organic farming systems. Farms with a variety of crops suffer fewer pest problems. Organic farms strive for diverse and balanced insect and plant populations. "Beneficial" insects and birds help control pests and also pollinate crops. Special plant/flower borders may be used to encourage beneficials. New Mexico's high percentage of small farms is advantageous for biodiversity and organic production.

Heirloom turkeys and chickens

Seed diversity, for example having many different varieties of corn, is also very important. Crop seeds have been developed over thousands of years, and have diverse abilities to grow under different conditions, like the hot, dry conditions of the Southwest. Heirloom, open-pollinated crops, an important feature of the organic garden, are passed along from one generation to another. Farmers have collected and planted their own seeds since farming began and seed collection can enhance crop suitability to local conditions. But many seed types have disappeared, along with a loss of plant variability. Hybrid and patented crop seeds have some advantages but they serve to reduce seed diversity and create dependence on a seed industry that is increasingly controlled by a few large corporations.

As with seed diversity, traditional livestock breeds are also disappearing. Most food animals raised today represent a relatively few livestock breeds developed for their ability to reach market size quickly. Heirloom or heritage breeds are maintained by dedicated individuals who believe in the value of diversity.

Readers who want to learn more can attend New Mexico's annual *Organic Farming Conference* held in February, usually in Albuquerque.

San Isidro Chapel

FAIRS, FESTIVALS, AND MUSEUMS

New Mexico celebrates its strong agricultural heritage through a variety of cultural institutions including religious celebrations, harvest festivals, fairs, and museums.

SAN ISIDRO, PATRON SAINT

San Isidro (also Ysidro) is the patron saint of farmers and farm workers. There are more than a dozen San Isidro churches in New Mexico and the saint is also revered in artwork. San Isidro Day is celebrated in the spring when a procession of children, dignitaries, community members, and the local priest carries a statute of San Isidro to the fields. There the priest blesses the fields and acequia, and asks for fertility, gentle rain, and protection from pests.

Bulto - San Isidro

STATE AND COUNTY FAIRS

Fairs celebrate rural life and offer a look at farm products of all kinds. New Mexico County Fairs showcase the best of local farms, including award winning farm animals, fruits, vegetables, woolens, quilts, jellies, pickles, and pies. County fairs are usually coordinated by Cooperative Extension.

Livestock entries
New Mexico State Fair

The biggest fair of all, the New Mexico State Fair, is held in Albuquerque in mid-September. Regional fairs and celebrations include the Southern New Mexico State Fair in Las Cruces, the Southwestern New Mexico State Fair in Deming, and the Eastern New Mexico State Fair in Roswell. Rodeos and livestock shows are held around the state. Pueblos and Tribes hold their own fairs, harvest festivals, corn dances, and feast days.

State Fair showcase

HARVEST FESTIVALS

Summer and fall harvest festivals and fairs present many opportunities to celebrate and taste local foods and regional specialties. The list below is hardly complete. Search out other harvest and food festivals in your local area. Check newspapers, magazines, web sites, and tourist offices for events and locations.

§ **Albuquerque Tomato Fiesta.** (August)
§ **Corrales Harvest Festival.** (October)
§ **Hatch Chile Festival.** (September, Labor Day Weekend)
§ **Harvest festival at El Rancho de las Golindrinas.** (October)
§ **High Rolls Cherry Festival.** (June)
§ **Corn mazes** and pumpkin fests statewide. (October)
§ **Local Food Festival and Field Day.** Albuquerque. (October)
§ **Mimbres Valley Harvest Festival.** (October)
§ **Pinto Bean Festival.** Moriarty. (October)
§ **Pueblo and Tribal Harvest Festivals.** (Fall)
§ **Urban Farm and Harvest Festival.** Albuquerque. (September)
§ **Wine festivals.** Statewide. (Spring and Fall)
§ **Wool Festival.** Taos. (October)

Corn maze

Museums

New Mexico Farm and Ranch Heritage Museum in Las Cruces is the premier museum for farming history. Educational exhibits showcase the stories and history of New Mexico farming and ranching, including the early Native farmers. Outdoor exhibits feature cow milking, livestock, blacksmithing, and gardens.

El Rancho de Las Golondrinas, south of Santa Fe, is a living history museum, with buildings that date to the early 1700s. It was the last stop on the Camino Real before Santa Fe. Actors clothed in the styles of the past demonstrate life in Spanish Colonial New Mexico such as wool dying, spinning and weaving, and soap making. Features include a water powered grist mill for grinding corn and wheat, fields of traditional crops, the acequia, and livestock pens.

The **B-Square Ranch** in Farmington celebrates farming tradition. This is a living farm, productive and diverse, with traditional and experimental crops and livestock along the San Juan River. Displays include antique horse-drawn farming implements plus a wildlife museum. An agricultural trust dedicates this land to farming permanently.

The **Chile Pepper Institute** is located in Las Cruces on the New Mexico State University campus. The institute has educational displays and a chile library, and sells books, posters, and chile seeds. Nearby at the Fabian Garcia Horticultural Farm, a chile teaching and demonstration garden features more than a hundred chile varieties.

El Camino Real International Heritage Center between Socorro and Truth or Consequences celebrates the original road from Mexico City to Santa Fe. Exhibits describe a 400-year history of colonization, trade, travel, and cultural exchange.

The **National Hispanic Cultural Center** in Albuquerque presents history and cultural exhibits. Outside the center is a sculptural celebration of the acequia tradition with water flowing through a canoa, an unusual and innovative method of water delivery.

Other museums and sites with exhibits and artifacts highlighting historic agriculture in New Mexico include the **Indian Pueblo Cultural Center** in Albuquerque, the **Museum of Indian Arts and Culture** in Santa Fe, Pueblo and Tribal museums and visitor centers around the state, and prehistoric ruins at Chaco Canyon, Bandelier, Aztec, and other places.

El Rancho de Las Golondrinas

B-Square Ranch

Canoa sculpture
National Hispanic Cultural Center

Santa Fe farmers market

FOOD SECURITY AND SUSTAINABILITY

Sustainability entails looking toward the future. We can have what we need tomorrow by acting wisely toward our resources today. Planning for food security includes issues such as access to food that is both healthy and affordable, protection of farmland and water resources, even seeds and livestock.

More than 90% of New Mexico's food is imported from other states or countries. Food may travel thousands of miles to our tables. Such long distances make it more difficult to protect the safety and reliability of the food supply chain. Even if food is plentiful, industrial-scale agriculture can pose problems like disease epidemics and pollution from wastes. These are some of the reasons why support for local farms is important. Taking pride in local production should also translate into a more healthy environment.

New Mexico is still very rural with functioning and productive farmland. It meets all the conditions for quality food production: good water, soil, and sunshine. Year-round growing seasons can be created with greenhouses and row covers. We could indeed grow most of our food if necessary. That would be food security. New Mexico's cultural mix, extending from the first corn farmers of a thousand years ago, to the crops and foods of later settlers, to the diversity of today, suggests that sustainability is workable.

Diversity and sustainability - New Mexico high country farm

Farm conservation easements and agricultural land trusts have begun to provide long-term protection for farmland. Other community efforts to protect local foods and farms, described in a previous chapter, support community health, local jobs, lower transportation costs, and fresher food.

Sustainability has become a topic in school curricula. The Camino de Paz School & Farm in Santa Cruz is a Montessori school for grades 7-9. It includes a community-supported farm (CSA) where students grow and sell organic vegetables using traditional farming methods, including working draft horses. They use fiber from their sheep and goats, run a goat dairy, and have a stand at the farmers market, learning about marketing, pricing, supply, and demand. Some small farms and CSAs have internship programs providing hands-on experience for people who want to learn about farming.

Working draft horse

NMSU farm

NMSU Campus - College of Agriculture

Spring backyard garden - Las Cruces

Training and degrees in agricultural sciences and skills are available at high schools and colleges. County Cooperative Extension offices provide information on farming, gardening, nutrition, and food preparation. New Mexico State University has twelve farms (agricultural science centers). These provide demonstrations and field days on farming techniques, sustainable farming, water conservation, crops, and livestock. NMSU also has many agriculture-related degree programs. Lastly, a nod to home gardeners who learn by doing, little by little, year by year, connecting to growing food and making their food supply a little more secure.

Lettuce field at sunrise

Top of the watershed - Sangre de Cristos

The fence

SELECTED READINGS:

Barrett, E. **The Spanish colonial settlement landscapes of New Mexico, 1598-1680.** UNM Press. 2012.

Bidal, L. **Pisacah:** a place of plenty. McKee. 1995.

Carlson, A. **The Spanish American homeland:** four centuries in New Mexico's Rio Arriba. Johns Hopkins U Press. 1990.

Clevenger, T. **A History of vegetable crops in New Mexico**. NMSU Ag Exp Station Bull 624. 1974.

Crawford, S. **River in winter.** "Acequia as Teacher" and "Open Letter to the State Engineer." UNM Press. 2003.

Davis, W. **El Gringo**: **New Mexico and her people**. Harper. 1856.

Dunmire, W. **Gardens of New Spain:** How Mediterranean plants and foods changed America. U Texas Press. 2004.

Dunmire, W. **New Mexico's Spanish Livestock Heritage**. UNM Press. 2013.

Ford, R. **Prehistoric food production in North America.** U Michigan. 1985.

Harris, R. **Artisan farming:** lessons, lore, and recipes. Gibbs Smith. 2008.

Lister, R. **Chaco Canyon:** archaeology and archaeologists. UNM Press. 1981.

Madison, D. **Local flavors:** cooking and eating from America's farmers markets. Broadway. 2002.

Matson, R. **The Origins of southwestern agriculture**. U Arizona Press. 1991.

Padilla, C. **The Chile chronicles**: tales of a New Mexico Harvest. Museum NM Press. 1997.

Price, V. **Canyon gardens:** the ancient Pueblo landscapes of the American Southwest. UNM Press. 2006.

Riley, C. **Rio del Norte:** people of the Upper Rio Grande from earliest times to the Pueblo revolt. U Utah Press. 1995.

Rivera, J. **Acequia culture:** water, land, and community in the Southwest. UNM Press. 1998.

Robinson, J. **The farmers market book:** growing food, cultivating community. Quarry. 2007.

Shafer, H. **Mimbres archaeology at the NAN ranch ruin**. UNM Press. 2003.

About the Author

Timothy ("Tim") McKimmie lives in Las Cruces. He has been involved with southwest agriculture, horticulture, and botany for more than 30 years, retiring in 2012 from the College of Agriculture at New Mexico State University where he also worked as agriculture librarian. He is passionate about New Mexico, its landscape, and its agricultural heritage. Other publications include *Chihuahuan Desert Gardens: a native plant selection guide*.

High country range

CREDITS

Author photo, page 102, by Dee Sorenson.

All other photographs by the author. All were taken in New Mexico.

Thanks to the New Mexico Farm and Ranch Heritage Museum for permission to use the photograph of their San Isidro carving (bulto), page 91.

Thanks to the Deming Luna Mimbres Museum for permission to photograph the Mimbres bowl, page 17 and the blanket, page 23.

Thanks to the New Mexico Wine Growers Association for permission to use the photograph of their Harvest Wine Festival poster, page 93.

Correspondence: unityswpress@gmail.com